W9-DDQ-097

FOSS Science Resources

Soils, Rocks, and Landforms

Full Option Science System
Developed at
The Lawrence Hall of Science,
University of California, Berkeley
Published and distributed by
Delta Education,
a member of the School Specialty Family

1487709
978-1-62571-348-3
Printing 7 – 5/2018
Quad/Graphics, Versailles, KY

Table of Contents

What Is Soil?

Have you ever dug a hole in the ground? What did you remove to make the hole? **Soil**. Sometimes people call it dirt, but a scientist calls the layer of diggable material that covers planet Earth soil.

What is soil? If you pick up a handful of soil and look at it closely you might be able to see and feel what soil is made of. Soil is mostly made of several sizes of **rock**. You might see pebbles and smaller pieces of gravel. Soil usually contains sand. **Particles** of sand are really tiny rocks. Some pieces of rock are even smaller than grains of sand. Smaller pieces are called **silt**. The smallest pieces of rock are clay particles. Clay particles are too small to see, but you can feel them. Clay feels slippery when it is wet.

So soil is a mixture of different-sized rocks (pebbles, gravel, particles of sand, and even smaller particles of silt and clay) along with water and air. Rocks, air, and water are **earth materials**. But there is more to soil than earth materials.

Digging into soil

Soil is mostly made of rock in several sizes along with water and air.

Rock Size Chart

Particle name	Average size
pebble	◯
gravel	ο
sand	∘
silt	·
clay	invisible to bare eye

Soil also contains organic material. Organic material is the remains of dead plants and animals. Plants send their roots into the soil and animals dig into the soil. When plants and animals die, their remains become part of the soil. Plants and animals **decay** into tiny pieces called **humus**. Humus provides **nutrients** for plants. Humus also helps the soil **retain** water.

What is an animal that lives in soil? Worms! Worms are good for soil and help plants grow. Worms burrow through the soil. As they move, worms mix the soil and make passageways for air and water. Worm waste also adds nutrients that are good for soil and plants.

A worm in soil enriched with humus

Not all soils are alike. Some kinds of soil have more humus. Some soils have more clay. Some have more sand, pebbles, and gravel.

Digging into Soils

1. What differences do you see in the soils shown above?

2. Where do you think these soils are found?

Weathering

Pebbles and sand are pieces of rock. Pebbles are pretty big. You can count a handful of pebbles. Pieces of sand are tiny. You can't count the particles in a handful of sand. All pebbles and sand particles start out as huge masses of rock the size of mountains. How do mountains break down into pebbles and sand?

The answer is **weathering**. Weathering is the breaking apart of rocks into smaller pieces. Weathering happens to all rocks when they are exposed to water and air.

Physical Weathering

Rocks break down in two ways. **Physical weathering** makes rocks smaller, but does not change the rocks in any other way. When a big rock falls from the side of a cliff, it breaks into lots of smaller rocks. All the **minerals** in the small rocks are the same as the minerals in the big rock.

When rocks get hot and then cold, they can crack. Sometimes water gets into cracks in rocks. Water expands when it freezes. It can expand enough to break big sections of rock along the crack. When ice melts, the rock may break into smaller pieces.

Physical weathering of cliffs

A rock weathered by freezing and thawing of water

Roots of trees and bushes can grow down into cracks in rocks. As roots grow, they make the cracks bigger. Sometimes the cracks get so big that the rock falls apart.

When rocks bang into one another, they get worn down. Rubbing, grinding, and banging is called **abrasion**. Abrasion is a kind of physical weathering. It happens when rocks fall in **landslides**, tumble in flowing water, or crash around in waves. Wind can blow sand against rocks. This sandblasting weathers the rocks.

Tree roots grow and break rocks.

Sand abrasion on cliffs

Sand carried by wind can weather rocks into interesting shapes.

Chemical Weathering

Chemical weathering happens when minerals in rocks are changed by chemicals in water and air. The starting minerals change into new substances.

Many rocks contain iron. When oxygen in air comes in contact with iron, the iron in the rock can rust. Rust is iron oxide. Iron oxide is softer than other iron minerals. This causes the rock to break apart faster.

Carbon dioxide gas in the air **dissolves** in water droplets. This makes **acid**. The acid droplets can fall as rain. The acid causes the **calcite** in **limestone** and **marble** to make holes. This is a chemical change. Monuments, buildings, and gravestones made of marble or limestone change and weaken when exposed to acid rain.

Salt can cause chemical weathering. Salt water can **react** with minerals in rocks to make new minerals. When the new substances are softer than the original mineral, holes can form. The weak rock breaks and falls apart more easily.

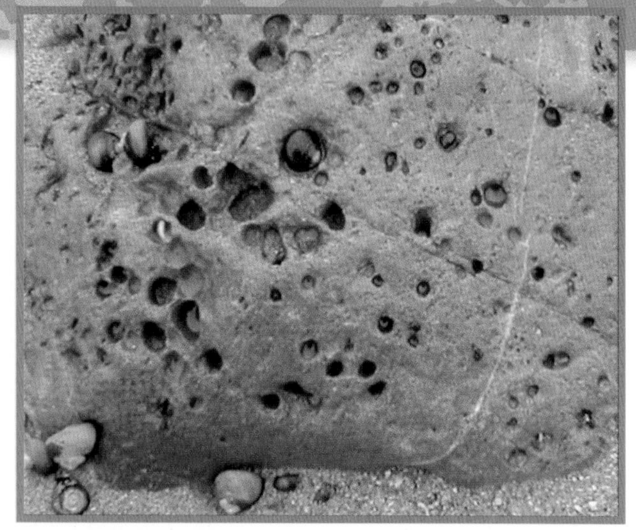

Chemical weathering of a rock containing iron

Chemical weathering of marble by acid rain

Chemical weathering of sandstone by salt water

Erosion and Deposition

A trip to the beach is fun. One of the best parts is playing in the sand. And there is so much sand. Where did it all come from? Was it made right there, or did it come from some other place?

Much of the sand on the beach came from mountains. **Erosion** moved the sand from the mountains to the beach. Erosion is the taking away of weathered rock. After rocks have weathered into small pieces, they can be carried away by gravity, water, or wind. Most of the sand shown here was carried to the beach by water flowing in rivers and streams.

As long as water keeps flowing, the bits of sand keep moving downstream. When the river enters the ocean, the water slows down. The sand settles to the bottom of the ocean. The settling of **sediments** is called **deposition**. Deposits of sand form beaches all over the world.

A beach

Erosion

The beach sand might start on high mountain cliffs. Sometimes big chunks of rock fall off the sides of mountains. Gravity pulls rocks downhill. Other times landslides move rocks and soil downhill.

Rainwater moving over the ground erodes the broken rocks. Water **transports** rocks into creeks. Water flowing in creeks transports broken rocks downstream. This process is called erosion.

Cliffs high in the mountains

Weathered rock in a mountain creek

Strong river currents move rocks downstream.

Creeks flow into rivers. Rivers have strong currents. Rivers can carry many sizes of rocks. The rocks bang together and rub on the riverbed. The rocks break into smaller and smaller pieces. The smaller pieces are pebbles, gravel, sand, and silt. Erosion continues. The farther the rocks move in the river, the smaller they get. They also get smoother and rounder as they tumble along.

Smooth, round pebbles along a river

Deposition

When the water flowing in a river slows down, the rocks are deposited as sediments. Large rocks are the first to settle to the bottom. Powerful **flood** waters move rocks of all sizes, even large boulders.

Where a river flows into a lake, a bay, or the ocean, the water slows down. Sand is deposited near the mouth of the river. The sand can form sandbars, deltas, and beaches. Farther out are deposits of silt and clay.

Large and small sediments deposited after a flood

Can you see deposits of sand and silt where this river enters the lake?

Can you see meanders in the river?

Other Kinds of Erosion and Deposition

Wind blows sand and smaller pieces of rock from one place to another. Sometimes the wind blows hard enough to carry a lot of sand and dust. Wind can erode valuable farmland.

When the wind dies down, sand and dust are deposited far from their starting places. This is how sand dunes form. Death Valley in California and Great Sand Dunes in Colorado are two places where large sand dunes formed.

Strong winds move earth materials from one place to another.

Sand dunes in Death Valley National Park, California

Great Sand Dunes National Park, Colorado

A U-shaped valley eroded by glaciers

Glaciers are frozen rivers. Rocks can be frozen in glaciers high in mountain canyons. Glaciers flow slowly through canyons. The frozen rocks scrape the floor and sides of the canyon. Glaciers weather and erode V-shaped canyons into U-shaped valleys.

Thousands of years ago in the Western United States, glaciers scraped down mountain valleys. They crushed and ground up rock beneath them. At the same time, glaciers covered much of the Midwest. These sheets of ice were over 1.5 kilometers (km) thick. They changed much of the landscape by eroding the surface and depositing the rock material in new places.

What happens when sand finally makes it to the ocean? Is that the end of the erosion and deposition story? Not quite. Waves erode beaches and deposit sand in different places all the time. As waves crash on the beach, sand continues to weather. Sand gets finer and finer. Sand abrades the rocks and cliffs along the ocean shore. Erosion and deposition go on and on.

Sand deposited on a beach around a weathered rock

Reviewing Erosion and Deposition

1. Describe and give examples of erosion.
2. Describe and give examples of deposition.

Landforms Photo Album

Landforms Formed by Weathering and Erosion

Arch A curved rock that forms when chemical and/or physical weathering weakens the center, and the rock erodes.

Arches can form on the land or near the coast where waves batter and erode the center of the rock.

Butte A hill with steep sides and a small, flat top. A butte is smaller than a mesa.

Mesa A single, wide, flat-topped hill having at least one steep side.

Gorge A narrow, steep-sided valley or canyon.

Valley A low area between mountains where a stream or glacier flows. Stream valleys are V-shaped. Glacier valleys are often U-shaped.

Hanging valley A valley floor above another valley floor. Glacial erosion causes hanging valleys.

Canyon A V-shaped gorge with steep sides eroded by a stream.

Meander A curve or loop in a river or stream.

Hoodoo A rock shaped like a mushroom or statue. Hoodoos are formed when weak rocks erode away and leave behind stronger rocks.

Exfoliation dome A dome formed when rocks like granite peel away at Earth's surface.

Spheroidal rocks Rounded rocks formed by physical and chemical weathering.

Landforms Formed by Deposition

Alluvial fan A fan-shaped deposit of rocks formed where a stream flows from a steep slope onto flatter land.

Beach An area made of sand and other sizes of rocks between the low-tide and high-tide levels at the coast or a lake.

Floodplain Land covered by water during a flood. Small particles, like sand and silt, are deposited on a floodplain.

Delta A fan-shaped deposit of earth materials at the mouth of a stream.

Sandbar A long ridge of sand in shallow water, built up by river currents or ocean waves.

Levee A bank along a stream that may stop land from flooding. Levees can be natural or made by people.

Moraine The unsorted rocks and soil carried and deposited by a glacier.

Outwash plain A flat or gently sloping surface made of sorted sediments deposited by water from melting glaciers.

Plain A low area of Earth's surface that is often formed by flat-lying sediments.

Sand dune The sand deposited by wind in ridges, mounds, or hills.

Landslide The rapid downslope movement of earth material.

Slump A downward movement of a single mass of earth material.

Landforms Formed by Eruptions

Volcano A place where lava, cinders, ash, and gases pour out through openings in Earth's surface.

Caldera A hole that forms when the top of a volcano blows off or when the magma below the volcano drains away.

Cinder cone A volcano formed from a pile of cinders and other volcanic material blown out in an explosive eruption.

Composite volcano A volcano built by alternating eruptions of lava, cinders, and ash. Mount Rainier and the other volcanoes in the state of Washington are composite volcanoes.

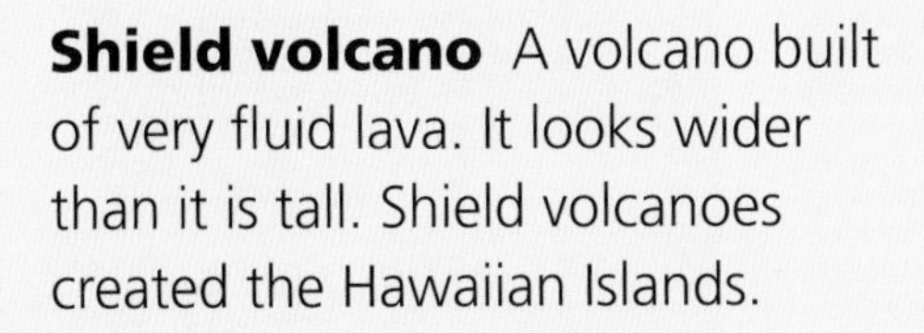

Shield volcano A volcano built of very fluid lava. It looks wider than it is tall. Shield volcanoes created the Hawaiian Islands.

Landforms Formed by Crust Movements

Fault A break in Earth's crust where blocks of rock **fracture** and move. The San Andreas Fault has created a wide crack in Earth's surface.

Plateau A high, nearly level, uplifted area composed of horizontal layers of rock. The Colorado River has eroded the Colorado Plateau, forming the Grand Canyon.

Mountain A high, steeply sloped area where rock is uplifted along a fault or created by a volcano.

Fossils Tell a Story

Earth scientists called **geologists** and **paleontologists** ask questions about the history of Earth. Geologists focus their studies on the structures of rocks and how they formed. Paleontologists ask questions about what life was like millions of years ago. These scientists can't travel back in time, so they look for answers to their questions in rocks. From the **evidence** they find, paleontologists build models to explain how they think plants and animals lived and what their world was like.

Digging through the Layers

Over billions of years, Earth's surface has changed dramatically. Areas once covered by water are now dry. Continents have moved around on the globe. Large areas of Earth's surface were once covered by thick sheets of ice, which have now melted.

Through all these changes, layers of new rock have formed on Earth. Some layers were formed by molten **lava** that erupted from **volcanoes**, covered Earth's surface, cooled, and turned into solid rock. Other layers of rock formed from tiny particles of rock called sediments that settled in Earth's ocean, lakes, and swamps. After millions of years these layers of sediments changed into **sedimentary rock**. By digging through layers of sedimentary rock, earth scientists have determined the age of Earth and discovered the **organisms** that lived here at the time the sediments were deposited.

Scientists dig through layers of soil and rock.

A fern fossil

Fossils Found

Scientists attempt to assign age to the rocks they study. They have several tools they use to determine the age of rocks. One technique is based on the **fossils** they find in them. Fossils are the remains of organisms preserved in rocks. Scientists know that some of the organisms represented by the fossils are much older than other fossils they find. This kind of information tells a scientist that one rock is older than another, based on the fossils found on them, but not how old either rock is in absolute terms. For determining the absolute age of rocks, geologists use other tools and techniques.

Fossils form in several different ways. Most form when a dead organism ends up at the bottom of a water environment (the ocean, a lake, or a swamp). The organism gets buried in mud and sediments. Most of the softer parts of the organism decompose, but the hard parts (bones, teeth, some shells) remain. Sediments carried by erosion continue to deposit on top of the remains. Over very long periods of time, the sediments are squeezed and compacted and eventually harden into rock. As buried bones age, minerals seep into the bones and bit by bit turn the bones into solid rock. A bone fossil is no longer bone but rock in the exact shape of the original bone. This is the same way a tree turns into **petrified wood**. A piece of petrified wood is a tree fossil, preserving the tree's wood grain and growth rings exactly.

A fish fossil

Sometimes a buried organism decomposes, leaving a space in the sediment where the organism was originally. Later, as the sediments continue to change into rock, the space in the sediments called a **mold** will fill with minerals. The minerals that fill the mold turn into a different kind of rock than the sediments surrounding the mold. After a long time, the minerals in the mold form an exact replica of the original organism. This kind of fossil is called a **cast**. Cast fossils of organisms, like clams, sea urchins, and snails, are fairly common.

Look at the mold and cast of an ammonite fossil. Ammonites are extinct ocean animals that had spiral shells. They are excellent index fossils and are used to date the rocks.

After years of studying fossils found all over the planet, paleontologists have discovered that some of the fossil organisms lived on Earth for only a short time, perhaps a few million years, and are very wide spread. A few are found all over the world, in the United States, Europe, Asia, and South America. Such short-lived, widespread fossils are known as index fossils. Index fossils help Earth scientists determine the absolute age of the rocks they are studying. When a scientist locates an index fossil in the rock layer she is studying, she knows a lot about that rock layer and when its sediments were deposited.

Fossilized tree trunks

The Fossil Record

Together, all the fossils on Earth make up the **fossil record**. The fossil record is a valuable source of information about the history of life on Earth.

The fossil record also provides evidence of the environment in which ancient organisms lived. An important part of an organism's environment is climate, the average weather in an area.

We have learned from fossils that Earth's climates have changed over time. For example, fossils of plants that would have thrived in hot and humid areas (tropical jungle) have been found in areas that are deserts now. Such fossil evidence suggests that these areas were wetter millions of years ago than they are today.

Fossilized tree trunks can provide similar kinds of evidence. Most trees do not grow as fast when the temperature is cold and dry. A petrified tree trunk with narrow growth rings suggests that the climate was probably cool and dry when the tree was living. A petrified tree trunk with wide growth rings suggests that the climate was probably warm and moist during the tree's lifetime.

Sometimes a scientist finds fossils of ocean animals high up on a mountain. What do you think she might conclude about the history of that area? Why?

Pieces of a Dinosaur Puzzle

Most dinosaur fossils date from 65 million to 225 million years ago. Fossils help scientists determine what dinosaurs looked like and how they lived. For example, a dinosaur's teeth reveal whether it ate meat or plants. The size of a limb can tell if the dinosaur walked on four legs or two.

When paleontologists find a complete skeleton, they can assemble it to show what a dinosaur looked like. The bones might come from more than one skeleton of the same type of dinosaur. They build a full model of the dinosaur's body based on the bones they have. They construct bones to replace the ones that are missing to make a complete skeleton. This process of filling in the gaps of missing bones is called a **restoration**. Many museums display dinosaur models that are restorations. Often the entire skeleton is not made of the real bones, but of copies of the original model.

A scientist assembling dinosaur bones

A dinosaur skeleton restoration

James Kirkland, a paleontologist and geologist, brushing off a fossil at a Utah site where scientists have found the fossils of hundreds of small dinosaurs

Paleontologists around the world continue to piece together models of dinosaurs based on newly unearthed fossils. New discoveries add information to our knowledge of past life on Earth.

One thing that paleontologists are interested in finding out is what dinosaurs ate. The earliest dinosaurs were small-bodied, fast-moving **predators** that hunted and ate other animals. They were **carnivores**, or meat eaters. But later, the fossil record shows that two other major groups of dinosaurs existed and these two groups were both **herbivores**, or plant eaters. How did these plant-eating dinosaurs evolve from the early meat eaters? This was a mystery, and there was no fossil evidence to provide the answers.

In 2002, paleontologists began digging bones out of rocks in east central Utah. In one location, they found fossils of hundreds of medium-sized dinosaurs with long necks and long, clawed hands. They compared the dinosaur skeletons to existing dinosaur bones. When the new bones did not match, the scientists came to an interesting **conclusion**. They had discovered a new type of dinosaur. They published their finding in 2005.

The new dinosaur was about 4 meters (m) long, head to tail. It stood about 1.4 m tall and might have weighed about 225–450 kilograms (kg). It had sharp, curved claws that were 10 centimeters (cm) long. With almost 1,700 bones excavated between 2002 and 2005, scientists have about 90 percent of the dinosaur's bones.

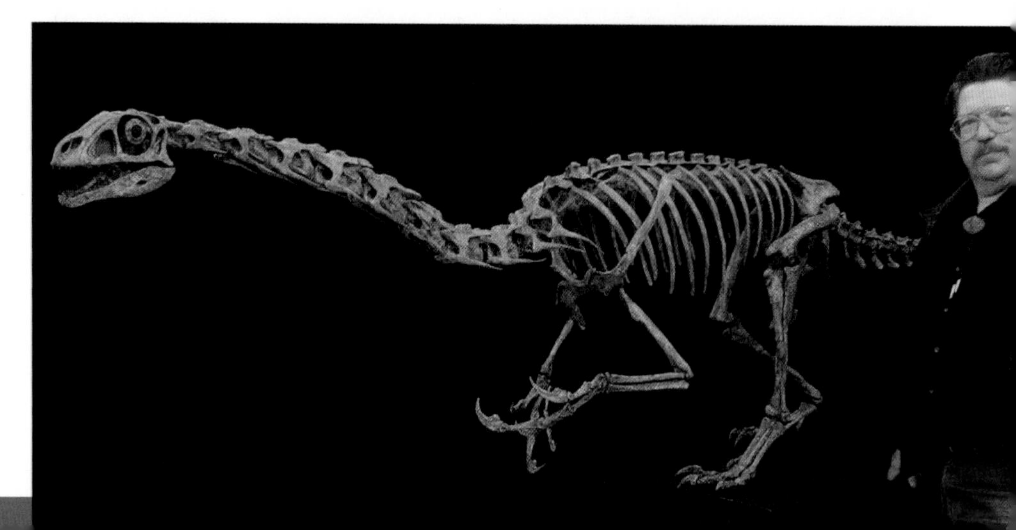

Kirkland standing near a full-size model of the newly discovered dinosaur with sharp, curved claws

They named the newly discovered creature *Falcarius* (pronounced fal-cah-RYE-us), Greek for "sickle bearer" because of its claws. Based on the rocks where it was found, it lived in the early Cretaceous period about 125 million years ago. The sharp, curved claws indicate it hunted and ate small animals like its dinosaur relatives.

One thing that was so interesting about this new dinosaur was its teeth. They were shaped like tiny leaves. This would be a good shape for shredding plants. Meat-eating dinosaur relatives had blade-like teeth. *Falcarius* was different from its relatives.

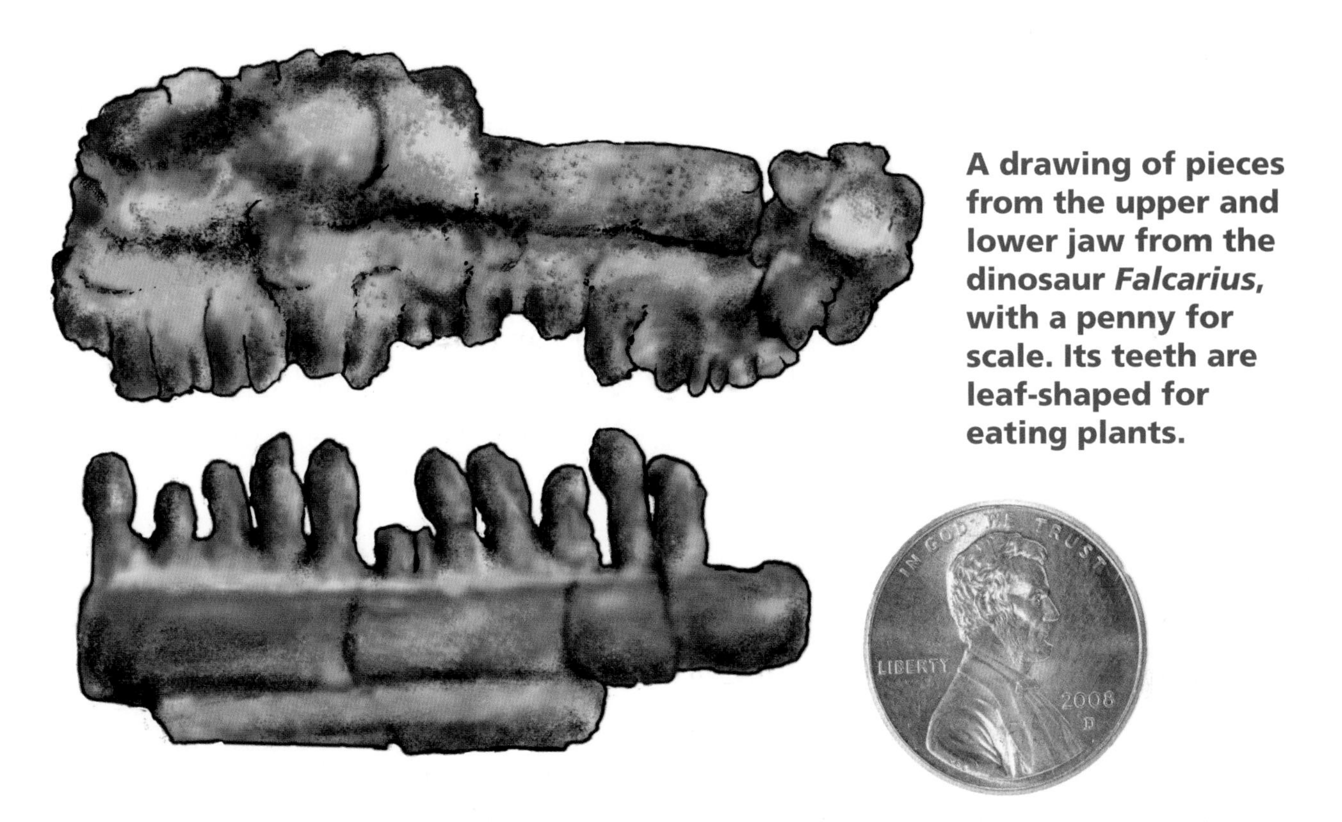

A drawing of pieces from the upper and lower jaw from the dinosaur *Falcarius,* with a penny for scale. Its teeth are leaf-shaped for eating plants.

Another interesting thing was the pelvic bone of *Falcarius*. The shape of that bone was broad, providing evidence that the dinosaur had a very large **digestive system**. A large gut would be needed if the animal ate plants.

The lower legs of *Falcarius* were stubby. It couldn't run very fast to catch prey. Compared with carnivorous relatives, *Falcarius's* neck was longer and its forelimbs were more flexible. These features made it easier for the dinosaur to reach plants to eat.

All together, the analysis of these bones revealed something extraordinary. This dinosaur was not just a carnivore like other dinosaur relatives that lived in the Cretaceous period. It also ate plants. It represented a stage between carnivores and herbivores.

During dinosaur evolution, two major groups of dinosaurs shifted to plant-based diets. But fossils of dinosaurs that were both meat eaters and plant eaters hadn't been found. Scientists didn't have the information they needed to understand the plant-eaters' relationship to their meat-eating ancestors. The *Falcarius* fossils show this transition in action among one group of dinosaurs. That group is the birdlike meat-eating and plant-eating dinosaurs of the Cretaceous period.

The discovery of *Falcarius* was a very important piece of the larger puzzle of dinosaur evolution. The dinosaur dig in Utah will continue to provide more information for paleontologists in years to come. The site has fossils of *Falcarius* babies, juveniles, and adults. Scientists will be able to compare the structural differences between young and adult dinosaurs, and between males and females. Scientists will be able to develop more models about how fast the animals grew, at what age they became adults, and how they lived in the environment so long ago.

An artist's conception of the dinosaur *Falcarius*

Topographic Maps

A map is a picture of Earth's surface. Although most maps are drawn on flat paper, Earth has many bumps and curves. These changes in elevation are difficult to show on a flat surface. A **topographic map** is a type of map that shows changes in elevation. At first, a topographic map may look like squiggles and blotches of color. However, the map can provide a great deal of information. First, you must understand how the lines and symbols are used.

The most obvious feature of a topographic map is the system of curved lines that covers the map. These lines are called **contour lines**, and they are usually brown. Each contour line represents a specific elevation. Elevations on US maps published by the US Geological Survey (USGS) are measured in feet above sea level. In countries where the metric system is used, elevations are measured in meters. Surveyors measure elevations when they gather the information to create a new map.

When you move your finger along a contour line on a topographic map, the elevation is the same at every point on the line. Your path would be flat if you walked along the course the contour represents in the real world. You would not go uphill or downhill. Contour lines are always connected at both ends. If you follow a contour line on a topographic map, you will always return to the place where you began. Sometimes the ends of a contour line run off the map page. But eventually they connect if you enlarge the map or follow them onto an adjacent map.

A hill on a topographic map is represented by rings of contour lines. The rings become smaller and smaller as they approach the top of the hill. If you move from one contour line to another, you move either up or down in elevation. In real life, if you walk across contour lines represented on the map, you will go either uphill or downhill.

Elevation changes rapidly if contour lines are close together. A steep hill is represented by closely spaced contour lines. If the contour lines are farther apart, they represent a more gradual slope. Some contour lines are thicker than others and have elevations printed on them. These are called **index contours**. The numbers tell the elevations of the index contours. Index contours help determine whether the elevation is rising or falling.

Topographic maps contain other important information. A scale is printed on the bottom of the map in the margin. In the United States, the standard scale of many topographic maps is 1:24,000 (1 inch equals 24,000 inches or 2,000 feet). The scale is shown as a ratio, graphic scale, or both. On some maps, the scale is given in meters. The text in the margin also states the **contour interval**. This is the change in elevation between any two contour lines. A contour interval of 3 meters (m) means that the elevation of each contour line is 3 m higher or lower than the one next to it.

Colors Used on Most USGS Topographic Maps

Green Major vegetation: forest, brush, and orchards

Blue Water: lakes, streams, rivers, springs, marshes, the ocean, and glaciers

Red Highways or boundaries

Black Human-made structures and place names

White Absence of vegetation: prairies, meadows, tundra, and deserts

Brown Land features, lava flows, sand areas, and contour lines

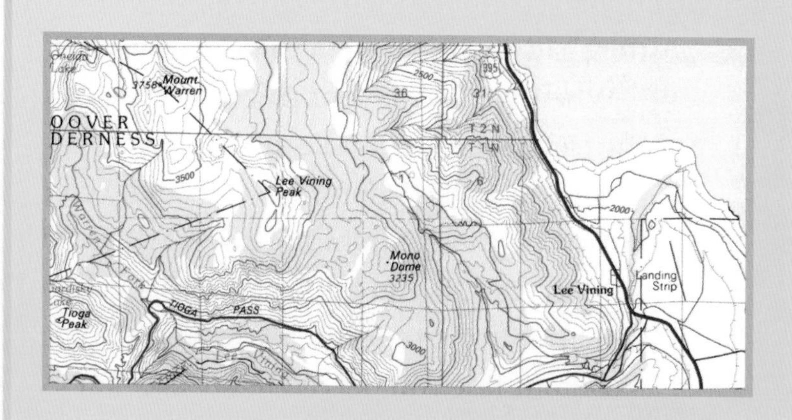

A topographic map of Yosemite Valley

Topographic maps contain much more information than changes in elevation. Symbols and colors represent natural **landforms**, vegetation, and structures made by people. USGS topographic maps use a standard set of colors and symbols to represent objects.

The United States Geological Survey (USGS)

Topographic maps are made by the United States Geological Survey (USGS). This federal agency was established in 1879. Its task is to map, study, and interpret the geology, hydrology, and topography of the country. USGS surveyors use the latest scientific instruments to create accurate maps. They have created topographic maps of the entire country. In addition, the USGS investigates natural hazards such as volcanoes, earthquakes, and landslides.

The Story of Mount Shasta

The following is adapted from an account of a night in a snowstorm on Mount Shasta. The story is based on an article by John Muir. The article was published in the September 1877 issue of Harper's New Monthly Magazine. *Muir (1838–1914) was a naturalist and explorer. He traveled extensively throughout California and wrote about his adventures.*

The climb to the top of Mount Shasta is usually undertaken in summer, during favorable weather. Then the deep snows have melted from the lower slopes. At that time, storms are much less likely to occur. But whatever the season, Shasta's peak is covered in a layer of snow and ice. I agreed to attempt the climb in spring in order to take barometric observations from the summit.

My companion, Jerome Fay, and I began our climb on April 30, 1875. We took packhorses to carry supplies. We made our way toward a camp about 16 kilometers (km) up the mountain trail. We planned to camp overnight and then get an early start the next morning.

A profile view of Mount Shasta

A sketch of Mount Shasta drawn by John Muir

We had not expected to encounter snow 1.5 meters (m) deep on the trail. We pushed on throughout the day, but made slow progress. As the Sun began to set, we realized we would not make our destination by nightfall. Determined to go on, we left the horses. We carried a day's provisions and our blankets up the slope to the timberline. There we set up camp, sheltered by a block of red lava. We slept only 2 hours. At 2:00 a.m., we arose to a fine, starry sky. After cooking our breakfast of venison on the coals, we set off for the summit.

Our pulses raced as we were surrounded by the beauty of the morning. We plunged ahead, hardly stopping for breath. Our boots clomped over the red lava apron that leads up the west side of the mountain. We made our way toward the smaller of the two cone-shaped summits. We crossed the gorge that separates the two peaks and swung around the Whitney Glacier.

The hot fumaroles, vents through which volcanic gases and steam escape, hissed and belched as we hiked upward. By 7:30 a.m., we were at the summit.

Fay and I marveled at the landscapes that surrounded us. I took the required barometric measurements and looked to my companion. At first, I did not understand his frown as he stared southward. As I looked into the Shasta Valley, I knew why he was dismayed. The valley was filled with gray and purple cumulus clouds. My first thought was how beautiful the storm clouds were on the mountain. Then I realized our danger.

Fay and I immediately began our descent. We moved slowly down the mountain. By 1:00 p.m. the storm reached the summit and began pounding us with hailstones. As we stood beside a hissing fumarole, I observed that the hailstones were of an unusual shape, with six straight sides and a domelike crown. Fay was interested less in the shape of the hailstones and more in the condition of the storm. All at once, the violence of the storm hit us with its might.

Had I been alone, I might have attempted the descent. However, Fay convinced me that the only sane course was to ride out the storm near the hissing hot springs. The temperature quickly dropped more than 11 degrees Celsius (°C). The wind became violent. Lightning flashed. We had only one chance. We removed as much clothing as possible and lay in the fuming mud on the edge of the hot springs. Thus we spent the night, freezing on one side and boiling on the other.

As we lay in this state, both threatened and protected by the mountain, my mind wandered to the origins of Shasta. The giant cone of Mount Shasta stands in constant snow. It can be seen from anywhere within a 80 to 160 km radius around the mountain. This majestic mountain of volcanic ashes and lava rises 4,317 m above sea level. It towers more than 3,000 m above the plain on which it sits. Shasta originated from repeated eruptions that built it upward and outward like the trunk of a tree. The mountain is more than 250,000 years old. Its two peaks are the results of multiple eruptions.

The remains of Shasta's violent history can still be found on the sides of the mountain. Gases, mud, steam, and boiling water spew up through cracks from the magma, or molten rock, below. The fumarole in which we were lying was one such outlet.

As quickly as it came upon us, the storm ended. The sky became clear and filled with stars. The temperature was still too cold to attempt our descent of the mountain, so we remained in our cooker until morning. Then we trudged down the mountain in our frozen, stiff clothes.

Luckily we were met at our camp by a friend with horses. As we descended, we could feel the Sun's warmth on our backs. I looked over my shoulder at the great white cone of Mount Shasta. The ordeal of the previous night seemed like a faraway dream.

A sketch by John Muir

A Cascades Volcano

Mount Shasta is located 65 km south of the California-Oregon border. At 4,317 m, it is the second-highest mountain in the Cascade Range. Mount Shasta is an example of a stratovolcano. Stratovolcanoes are volcanoes composed of lava flows, ash, and other material blown out by explosive activity. Shastina, the prominent cone on the west flank of Shasta, was active about 9,200 years ago. The Hotlum cone at Shasta's peak was active only a few centuries ago.

It Happened So Fast!

Some landforms are so old, you might think they've always been that way. The Sierra Nevada has been uplifting for more than 2 million years, and it continues today. The Colorado River continues to carve the Grand Canyon, as it has for more than 5 million years. Rock in the Appalachian Mountains began folding more than 480 million years ago. Most changes to Earth's surface are so slow, we can't see them happen.

But sometimes changes happen rapidly. Rapid changes affect people and landforms. Here are some examples of fast changes to Earth's surface.

Yosemite Floods of 1997

Floods caused a lot of damage in northern California in 1997. Three factors created the floods. There was deep snow in the mountains, warm temperatures, and heavy rain.

Yosemite National Park was hit hard. The Merced River, which flows through Yosemite Valley, rose higher than ever before. Water spread out and covered much of the valley.

Flood water in Yosemite

A sign showing the water level during the flood

A Yosemite hiking trail after the flood

In some places, the flood water was 3 meters (m) deep. Campsites were washed away. Housing for the people who worked in Yosemite was destroyed. When the water slowed, sand and other sediment were deposited all over the valley floor. The course of the river was different. Flood water has the power to change the land rapidly.

Water flowing out of Canyon Lake over the spillway at the height of the flood

Canyon Lake Flood of 2002

In the 1960s, a dam was built on the Guadalupe River to prevent flash flooding and to provide a water supply for central Texas. The Canyon Dam created a reservoir called Canyon Lake. The lake became a popular recreation area in the Texas Hill Country between San Antonio and Austin. This area of Texas is known worldwide for its potential for flash floods. In fact, it is part of an area called Flash Flood Alley.

In early July 2002, it began to rain heavily in the upper part of the Guadalupe River watershed. About 1 m of rain fell during 1 week, and it continued to rain. The runoff flowed into the river and down to Canyon Lake. The lake was already full because of the rain falling on it. The water began to flow over the spillway of the dam. A wall of water over 2 m high and 380 m wide went over the dam. For the next 6 weeks, an amount of water equal to one and a half times the amount of water in the lake flowed over the spillway and into a narrow valley behind Canyon Dam. This torrent of water cut into the valley floor. It washed away the oak trees, mesquite, and topsoil and carried tons of sediment downstream.

When the water stopped flowing, the valley became a gorge. The gorge is 1.6 kilometers (km) long, hundreds of meters wide, and up to 15 m or more deep. The new walls of the gorge exposed limestone rocks dating back 100 million years. Those rocks contain fossils of worms and crustaceans, and tracks of ancient insects and dinosaurs. Teams of scientists are carefully observing the area and recording the evidence they find. The limestone is very brittle and breaks away from the canyon walls.

The other feature that was exposed by the flood was an **earthquake** fault. The **fault** was known to be there before the flood, and now 800 m of the Hidden Valley Fault is visible. The gorge has become a fresh, new laboratory for geologists to study faults found in limestone.

The Gorge Preservation Society (GPS) is a local citizen's group that works with the Guadalupe–Blanco River Authority (GBRA) and the U.S. Army Corps of Engineers to protect and study the gorge. They lead public hikes to some parts of the gorge.

The Canyon Lake flood formed this gorge.

Big Thompson Canyon Flood of 1976

The state of Colorado was celebrating its centennial on July 31, 1976. People were enjoying their summer vacations camping along the Big Thompson Canyon. The canyon is northwest of Denver. The town of Estes Park is at the western end of the canyon.

Thunderstorms often occur in the Rocky Mountains, especially in the afternoon. On this day, a thunderstorm formed over the western end of Big Thompson Canyon and didn't move. It dumped over 30 centimeters (cm) of rain in less than 4 hours. The canyon is steep and narrow, and there is little soil to retain the water. By 9:00 p.m. that evening, a wall of water more than 6 m high roared down the canyon. It was a flash flood! It sped down the canyon at about 6 m per second. Huge boulders were swept down the canyon by the wall of water.

People in the lower parts of the canyon had no warning. Because the flood happened so fast, the only way for people to escape was to climb to higher ground in the canyon. Many people didn't have time to get out of the way of the water. There were 145 deaths. A lot of things were destroyed, including 400 cars, 418 houses, and 52 businesses. Most of the main highway along the canyon was washed out. More than 800 people were evacuated by helicopter the next morning. The Big Thompson Canyon Flood was one of the deadliest flash floods ever reported in the United States.

Since this event in 1976, early-warning systems have been put in place. This advance notice about possible flash floods helps people move to safety.

Many houses were destroyed in the Big Thompson Canyon Flood.

Mount St. Helens Eruption of 1980

Mount St. Helens is a volcano in the state of Washington. It is part of a chain of volcanoes called the Cascade Range. On March 20, 1980, an earthquake happened. The north side of the mountain started to bulge.

Two months later, with little warning, another earthquake happened. On May 18, at 8:32 a.m., a magnitude 5.1 earthquake shook Mount St. Helens.

The bulge disappeared as a large avalanche of rocky debris slid down the side of the volcano. **Pumice** and ash erupted.

The debris filled the valley below Mount St. Helens. Trees toppled over like toothpicks. Over 600 square km were flattened. Volcanic mudflows called lahars spilled into the rivers. There were 57 people killed. An estimated 12 million fish at a hatchery, and 7,000 deer, elk, and bears were also killed. The eruption destroyed or damaged over 200 homes, 27 bridges, 298 km of highway, and 24 km of railways.

Mount St. Helens after the 1980 eruption

The May 18 eruption lasted more than 9 hours. The plume of ash reached to over 20 km above sea level. It moved eastward at an average speed of 100 km per hour. Ash traveled as far as Idaho by noon the next day. It was also found on tops of cars and roofs in Edmonton, Alberta, Canada, the next morning.

Since 1980, Mount St. Helens has had several smaller eruptions. Scientists continue to carefully monitor the mountain to learn more about volcanoes and to warn people if they think another eruption is about to happen.

A view of the ash in Connell, Washington, after the eruption. Ash fell over 57,000 square km.

Two scientists standing by fallen trees in Smith Creek valley

Homes and highways damaged in the Northridge Earthquake

Northridge Earthquake of 1994

At 4:30 a.m., on January 17, 1994, people living in the Los Angeles, California, area got a jolt. It was an earthquake deep under the city of Northridge, California. The earth shook for 15 seconds. The magnitude of the earthquake was 6.7.

Earth's **crust** has a lot of cracks. The cracks are called faults. Earthquakes happen when huge sections of Earth's crust slide past each other on a fault.

The Northridge Earthquake happened on a fault geologists didn't know about. It was a blind thrust fault. Blind thrust faults don't reach all the way up to Earth's surface. They are hidden faults.

Damage was widespread. Sections of major highways fell. Parking structures and office buildings fell apart. Many apartment buildings were beyond repair. Houses in the towns of San Fernando and Santa Monica were also damaged. About 22,000 people lost their homes.

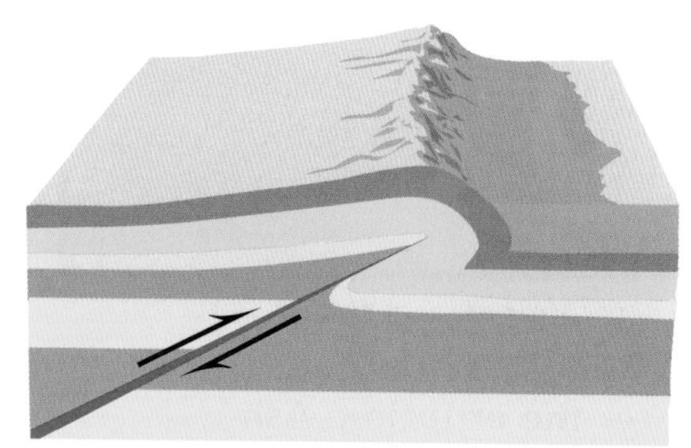

A diagram of a blind thrust fault

San Francisco soon after the 1906 earthquake, with smoke rising in the background

San Francisco Earthquake of 1906

Wednesday, April 18, 1906, was the day a big earthquake struck San Francisco, California. People felt the first little shakes at 5:12 a.m. Soon after, the major shaking started. It lasted 47 seconds. People as far away as southern California, Oregon, and central Nevada felt it. The magnitude of the earthquake was about 7.9.

Movement on the San Andreas Fault caused the 1906 earthquake. The fault broke at Earth's surface for a distance of 470 km. Cracks opened and cliffs formed where sections of land fell. Land near the San Francisco Bay settled as a result of the shaking. The settling and shaking caused buildings to fall. Great fires broke out because gas pipes broke. And the fires burned because there was no water to put them out due to water pipe damage.

Earthquakes usually last only a few seconds. Scientists can't predict earthquakes yet, but **engineers** can help people build stronger structures to minimize the damage.

The town of La Conchita at the base of an unstable hill

La Conchita Landslides of 1995 and 2005

Landslides occur when rocks and soil quickly slide downhill. Some areas are more likely to have landslides. The hillside above La Conchita, California, is one of those areas. This small town has had two large landslides. The slides killed people and damaged buildings and cars.

The landslide shown here happened on March 4, 1995. Many people were evacuated because of the slide. Houses nearest the landslide were completely destroyed. No one was killed or injured.

People continued to live in the area below the hillside. Another landslide happened on January 10, 2005. It destroyed or damaged 36 houses and killed 10 people. This kind of land movement happens so quickly, it is often impossible to get out of the way.

This landslide is another example of erosion and deposition. You can see where the sand and mud eroded from the top of the hill. You can also see where the sediment was deposited on the edge of town.

Yosemite Rockfall of 1996

On July 10, 1996, Ernie Milan was jogging on a trail in Yosemite National Park. Ernie was a trail worker for the National Park Service, so he knew the area well. He heard a loud boom. Dust started swirling around him. Day turned into night. What happened?

A giant mass of **granite**, weighing nearly 70,000 tons, broke loose from a cliff. It fell 600 m to the valley floor. Hundreds of trees were knocked over. One person was killed, and several others were injured. Ernie was not hurt.

Scientists estimated that the rock hit the floor at 400 km per hour. There is no way to stop rockfalls or predict when they will happen again.

Rockfalls happen where loose or cracked rocks are on steep slopes. Rockfalls may happen along road cuts and other excavations. Rockfalls start when rocks are dislodged by freezing or thawing of water or by heavy rainfall. Rockfalls can also be triggered by ground shaking from earthquakes. They generally occur without warning.

Rockfalls happen all over the world. They are a natural kind of weathering and erosion. Most rockfalls are not observed by people. But scientists try to learn what they can from these rockfalls. Some day scientists might be able to predict when a mass of rock is ready to break away.

A 70,000-ton mass of granite fell to the floor of Yosemite Valley.

Monumental Rocks

Humans build monuments to honor important people and events. Monuments are built to last a long time. They are usually large structures. If you were going to build a monument, what would you make it out of? Rock would be a good choice.

From ancient times to the present, people have made monuments out of rock. Why? Because rock is found everywhere. Rock can be cut and shaped. And most of all, rock lasts a long time. Some rocks are nearly as old as Earth itself. Some structures made of rock have been standing for thousands of years.

The Great Pyramid and Sphinx

The Great Pyramid

Did you know that the Great Pyramid in Egypt is almost completely solid? The only spaces inside are a few hallways and rooms. The pyramid is made out of about 2,300,000 blocks of limestone and granite! The average block weighs about as much as two cars. The largest blocks weigh as much as six cars.

As big as it is, the Great Pyramid was made to honor just one person. It was built around 2700 BCE to hold the body of the pharaoh Khufu. He was a ruler of ancient Egypt.

Each building block in the pyramid was pulled to the site on a wooden sled. Workers used copper axes, chisels, and saws to cut and fit the stones. Today, people wonder how such a building was made without iron tools.

Limestone for the pyramid's center was cut from nearby cliffs. That way, the stone did not have to be moved very far. Granite for the walls and doorways came from almost 1,000 kilometers (km) up the Nile River. Nicer-looking limestone for the outside of the pyramid came from a few kilometers away. By using barges, they floated the limestone down the river to the building site. Today, the polished limestone shell is gone. The stone was "recycled" in the 1300s to rebuild a city damaged by earthquakes.

What Is Limestone?

Limestone is a sedimentary rock that forms from calcium carbonate. Limestone forms under water. Tiny bits of calcium carbonate, some from shells of organisms, drift to the bottom of the ocean or bays. These pieces of calcium carbonate pile up for millions of years. The layer of calcium carbonate gets thicker and thicker. After a very long time, the bits of calcium carbonate turn into limestone.

The Taj Mahal

The Taj Mahal in Agra, India, is one of the wonders of the world. Many agree that it is the most beautiful building of all time. The Taj Mahal's designer was Shah Jahan (1592–1666). He built the monument to honor his wife, Mumtaz Mahal (1593–1631).

The Taj Mahal, which means "Crown Palace," is made entirely of white marble. Builders from all over the Middle East worked 22 years to make it. Inside the Taj Mahal, colorful marble was cut and pieced together like a puzzle. Forty-three different kinds of gemstones were used for decoration.

What Is Marble?

Marble is a metamorphic rock. Metamorphic rocks form when one kind of rock changes into another kind of rock. This usually happens when heat and pressure act on a rock for a long time.

Marble starts out as limestone. When limestone gets buried deep inside Earth, the pressure builds. The temperature goes up. After millions of years, the limestone changes into marble.

The Vietnam Veterans Memorial

The Taj Mahal and the Great Pyramid each honor a single person. The Vietnam Veterans Memorial was built to honor all the Americans who died in the Vietnam War (1955–1975). It was the idea of Vietnam veteran Jan Scruggs.

A competition was held to design the memorial. Maya Lin's (1959–) plan was chosen. At the time, Lin was 21 years old. She was a student at Yale University. Lin designed the monument as a black granite wall. The wall forms a V. She wanted the rock to rise out of the ground like two arms to embrace people. One arm points to the Washington Monument. The other arm points to the Lincoln Memorial.

The wall was finished in 1982. The names of more than 58,000 military men and women are written on the wall. Many people didn't like Lin's design at the time. But the memorial she designed is one of the most visited sites in Washington, DC.

Each year millions of people visit the Vietnam Veterans Memorial, designed by Maya Lin (right).

What Is Granite?

Granite is an igneous rock. That means it started as melted rock deep under Earth's surface. As the melted rock moved toward the surface, it cooled and crystallized. When you look closely at granite, you can see the crystals of the different minerals.

There are only a few places in the world where black granite is found. The beautiful black granite in the Vietnam Veterans Memorial comes from India.

The Washington Monument

George Washington died in 1799 at his home in Virginia. That year, Congress voted to move his body to the capital city. They wanted to bury it under a marble monument. But Washington's body was never moved, and 85 years passed before a monument was completed. During that time, politicians fought. There were problems raising money for the monument. Sometimes there weren't enough railway cars to deliver the marble. For 2 decades, work stopped completely.

After the Civil War (1861–1865), interest in the monument rose again. People were worried about it. They said the base was not strong enough to support the finished building. Some of the marble blocks were splitting. The original plans had been lost. Many thought the monument was ugly. Some wanted to knock it down and start over. Others wanted a new design. Finally work went on after the foundation was made stronger. The upper two-thirds of the structure was built.

The outside of the monument was constructed from marble taken from quarries in Maryland. At first, the two sections looked the same. But over time wind and rain have caused the marble sections to weather differently. Can you see a color difference between the original marble from one quarry and the later marble from a different quarry?

The great stone monument was finally completed in 1884. It is 169.3 meters (m) tall, which makes it the tallest structure in Washington, DC. It is also the tallest stone structure in the world!

Geoscientists at Work

Where do **geoscientists** go to do their work? The answer is, just about everywhere. That's because the prefix *geo–* means Earth. Geoscientists observe, investigate, and test landforms all over Earth. Their job is to discover, manage, and protect Earth's natural systems. Earth's rocks, minerals, soils, water, air, plants, animals, and fossil fuels together are called **natural resources**. Geoscientists study the history, distribution, use, and conservation of Earth's limited, valuable, nonliving natural resources.

Most geoscientists spend a lot of time doing fieldwork. That means they are outside in direct contact with Earth. Most geoscientists have a specialty. They focus on one part of the Earth system. And they use instruments designed by and with engineers to collect data about the Earth system they study.

Marine Geologists

Marine geologists study the ocean floor. They also study the boundary between the ocean floor and the continents, including continental shelves, estuaries, and bays.

Of course, landforms on Earth exist on land, but they can also be found under the water. Mountains, valleys, volcanoes, islands, plains, and canyons all exist in the ocean. In fact, Earth's highest peaks, deepest valleys, and largest flat plains are all in the ocean.

Marine geologists work with acoustic engineers to develop sonar instruments that allow them to "see" underwater structures by bouncing sound waves off rock formations. Sonar can show the size and location of volcanoes and canyons in deep water. Marine geologists use these data to map the bottom of the ocean, one of the most geologically active regions on Earth.

Atmospheric Scientists

Atmospheric scientists study the composition and activities of Earth's air. They may study weather and the effects of solar radiation on the atmosphere. They may also study atmospheric chemistry, including air pollution, global climate dynamics, and climate change.

Atmospheric scientists depend on a whole family of engineers to develop the instrumentation (weather balloons, gas samplers, aircraft, spacecraft, and satellites) to help them gather the data they need to learn about atmospheric phenomena.

Seismologists

Seismologists are the earth scientists who study earthquakes. They use seismographs and computer programs to find and monitor fault lines. They place motion-sensing instruments along known faults to collect data. These data help them develop systems to alert people when large, destructive earthquakes are about to happen.

Structural Geologists

Structural geologists record the shape of Earth's surface. They also study the massive forces that produce earthquakes and create mountains. They use maps and computer programs to analyze changes to Earth's surface.

Hydrologists

Hydrologists inventory and monitor Earth's fresh water. They measure the quantity and quality of drinking water in lakes, rivers, and wells. They monitor the amount and speed of water flowing in rivers and streams to anticipate the risk of flooding and soil erosion. Hydrologists use tools such as flow meters, depth meters, pH meters, and chemical tests.

Petroleum Geologists

Petroleum geologists explore Earth for deposits of oil and natural gas. They attempt to find the particular rock formations deep underground that contain these resources to produce petroleum. To do so, they send powerful sound waves into Earth and analyze the reflected sound waves using computer programs. When promising rock formations are detected, engineers then drill down to see if there is actually oil and natural gas. If they find deposits, the resources can be extracted and used as an energy source.

Volcanologists

Volcanologists study volcanoes and possible volcanic regions. They use tiltmeters, seismometers, and computers to determine when and where volcanoes might erupt. Volcanologists also study areas where volcanic activity is occurring. They determine if geothermal engineers might tap this heat source to generate electricity.

Soil Scientists

Soil scientists study the composition and quality of soils. They look for ways to keep farm soils stable and fertile for growing food crops. Soil engineers design field layout and maintenance plans to prevent soil erosion.

Corn and alfalfa crops are planted in alternating rows to protect them from soil erosion.

A cotton field with grass planted in rows to prevent wind erosion of the soil

A soil scientist field mapping soils using GPS technology

Soil Science at Home

Many home gardeners improve their garden soil with compost. Composting is a way to produce humus. With a little simple home engineering, you can make a compost bin. Toss in vegetable waste from the kitchen, lawn and plant clippings, dead leaves, and other plant material. As bacteria and fungi decompose the organic material, it slowly changes into pure dark brown humus. The humus can be worked into soil to enrich its fertility.

Kitchen waste and leaves are collected and broken down in a compost bin.

After several months, the compost has turned into humus and is added to garden soil.

A home gardener uses compost in garden beds to enrich the soil with humus.

Garden plants growing in soil with humus are large and healthy.

Sand, gravel, and pebbles are the aggregates used in concrete.

Making Concrete

Concrete is a rocklike construction material that is made by people. You have probably seen a lot of places where **concrete** is used. Many highways are made of concrete. Bridges and overpasses are often made of concrete. Dams and stadiums are made of concrete. What is concrete? You might see cement trucks at construction sites. They should be called concrete trucks because the material they carry in the big, round container is actually concrete. Most modern concrete is a mixture of **Portland cement** and aggregates. Aggregates are pieces of rock of different sizes. Small aggregates include sand and gravel. Larger aggregates include pebbles of several sizes.

Portland cement is a fine gray powder made from limestone. Limestone is dug out of a quarry. Then it is heated to a high temperature in a furnace and ground into a fine powder. When Portland cement is mixed with water, it makes a sticky, mudlike mixture. Over time, the cement mixture cures (hardens). The mixture changes into a solid, hard lump. Cured cement is as hard as rock.

In order to take full advantage of this **property** of cement, aggregates are added to the sticky mixture. The cement bonds to the pieces of rock and sand, cementing them together into one strong mass. The mixture of cement, water, and aggregates is a thick fluid that can be poured into forms. The big container on the back of the cement truck is always turning around and around. The motion keeps the concrete moving around so it won't harden inside the truck.

Cement trucks usually don't travel very far from the plant where they load up with ingredients. Many contractors require that the concrete be poured within 90 minutes after loading. If the concrete hardens in the truck, it might be necessary to use jackhammers to break up the concrete. Foundations for buildings are made by pouring concrete into forms. When the concrete foundation is cured, the building is constructed on top of the foundation.

Cement trucks deliver concrete to construction sites.

Concrete is poured into forms to make foundations for buildings.

The Roman Colosseum

The use of concrete for roads, buildings, bridges, and wharfs is not new. Concrete was invented and used widely by Romans more than 2,000 years ago. The Roman Colosseum, built in the year 70 BCE, is the largest colosseum in the world and is still standing. Concrete harbors built about the same time in the Bay of Pozzuoli near Naples are still strong. They have withstood battering from ocean waves and tides. They have endured countless earthquakes for 2,000 years. Modern concrete harbors constructed using Portland cement concrete last as few as 50 years. Why the big difference?

The difference is how the concrete is made. A team of scientists and engineers led by professors of civil and environmental engineering at the University of California, Berkeley, studied the concrete in these old structures. In 2013, the team leaders, Marie Jackson and Paulo Monteiro, announced the results of their analysis of the Roman concrete.

They knew from historical records that Roman-engineered concrete contained slightly different materials than modern Portland cement concrete. Roman concrete used baked limestone (like Portland cement). But it also used locally available volcanic ash, called pozzolan. The pozzolan seems to be the secret ingredient that makes the Roman concrete so strong and durable.

The wharf area in the Bay of Pozzuoli

The researchers also discovered that the Romans converted the limestone into lime by heating it to a much lower temperature than the 1,450 degrees Celsius (°C) needed to produce Portland cement. The cooking part of the cement-making process is important. It is important because it takes more fuel to generate the heat to convert limestone into Portland cement. The greater the amount of fuel burned, the greater the amount of greenhouse gas (CO_2) produced as a waste product. Using the Roman formula for converting limestone to lime requires less fuel, resulting in less CO_2 released into the atmosphere.

The potential engineering advantages presented by these discoveries are huge. The new way for making concrete can produce a stronger building material. Structural engineers can use it to design safer, stronger buildings, bridges, wharfs, and roads. Concrete manufacturers will be able to use much less energy to convert limestone into concrete-grade lime. The Roman concrete recipe will save energy and reduce the amount of CO_2 going into the atmosphere. And large deposits of pozzolan are available around the world.

Concrete foundations are different in different regions. For example, concrete in North Carolina is different from concrete in Texas, Wisconsin, or Oregon. That's because the aggregates mixed with the cement are always from the local region. It is too expensive to transport sand, gravel, and pebbles over long distances to make concrete, so the aggregates are local. Where did the aggregates come from to make the concrete foundation used to build your school?

Concrete can be used as stepping stones in a garden.

This school has concrete steps and walls.

Cutting the clay with a wire

Rolling the clay flat with a rolling pin

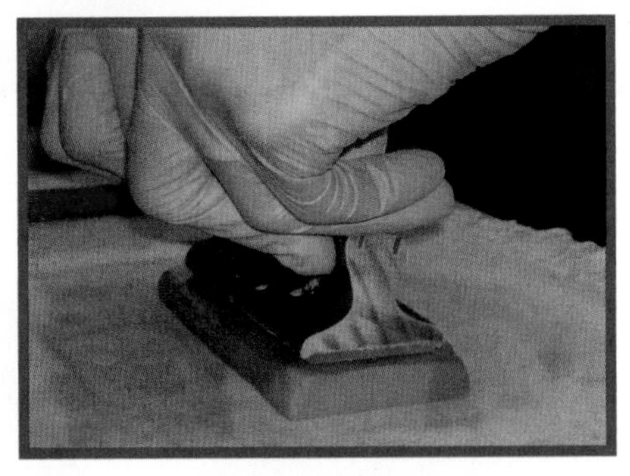

Using a stamp to press a shape into the clay

Earth Materials in Art

Rose Craig (1943–) is an artist. Rose makes beautiful watercolor paintings, and she is a skilled graphic artist and illustrator. For many years, Rose worked for the FOSS science program, drawing illustrations for articles in the *Science Resources* books. Rose has one other artistic interest, too. Rose makes ceramic tiles. Ceramic tiles are made of clay. As you know, clay is an earth material, so Rose is an earth material artist.

This is how she makes her beautiful tiles. First, she places a big pile of clay on her worktable. She uses a big rolling pin to flatten the clay into one big, thin sheet. The rolling process is very similar to rolling pie dough for a pie crust. When the clay is rolled out just right, Rose uses a straightedge and a knife to cut the slab of clay into rectangular pieces. Rose then uses rubber stamps to press a shape into the surface of the soft clay. She might use a fish shape, a flower shape, or a dragonfly shape, and then trim around the shape. Now Rose has to wait for the clay to dry.

A fish shape stamped in the clay

A kiln is a very hot oven that changes the clay into ceramic tiles.

Finished tiles

After 3 or 4 days, the tiles are dry and ready to be decorated with special paints called glazes. First, Rose paints a background color on the tiles. When that is dry, she presses the design into the clay again and enhances it with bright contrasting colors.

When the tiles are painted just the way Rose likes them, she puts them in a kiln. A kiln is an oven that gets really hot, much hotter than a pizza oven. The intense heat changes the clay into rock-hard ceramic tiles. The colored glaze becomes intensely shiny and hard as glass. The finished product is beautiful and useful. Because ceramic tiles are waterproof, they are good surfaces for sinks and counters that get wet. They are also useful outside in the garden or on a deck because sunshine, rain, or snow will not damage them.

Where Do Rocks Come From?

Where do rocks come from? This question keeps geologists busy. Even though they don't have all the answers, they know a lot about where rocks come from.

Earth is about 4.6 billion years old. The oldest dated rock found on Earth is about 4 billion years old. That's almost as old as Earth. Scientists have also found crystals of a mineral called zircon that were formed 4.4 billion years ago.

There are three big groups of rocks: **igneous**, sedimentary, and **metamorphic**. All the rocks in a group have similar origins, often inside Earth.

Earth is like an egg. An egg has a hard outer layer called the shell. Earth has a hard outer layer called the crust. Earth's crust is made of solid rock.

Under an egg's shell is the fluid egg white. Under Earth's solid crust is the **mantle**, partly melted rock that flows like really thick toothpaste. It is hot inside Earth. It is so hot that rocks and minerals melt.

An egg has a yolk in the center. Earth has a metal **core** in its center. Earth's core is made of iron and nickel.

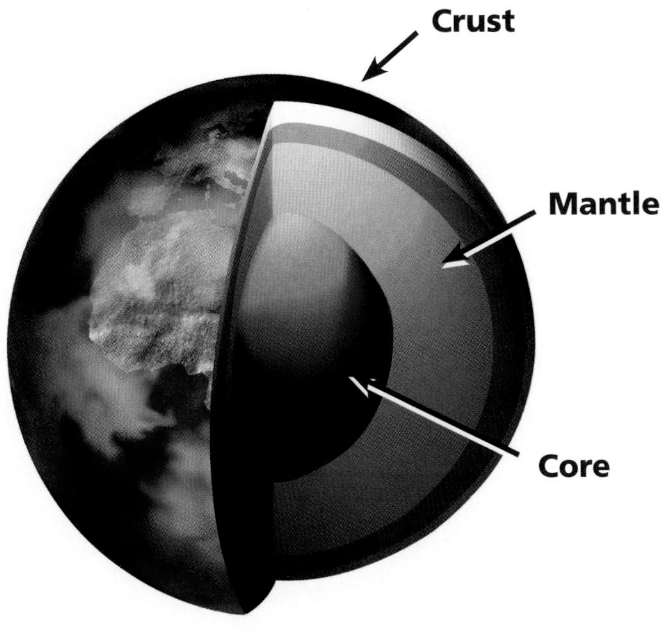

A cross section of Earth

Igneous Rocks

Igneous rocks start out as melted rock deep in Earth's crust. Sometimes the melted rock, called **magma**, comes to the surface in volcanoes. The magma pours out as lava. When lava cools and hardens, it forms new rocks. The basalt you tested for calcite is volcanic igneous rock. Much of the rock in the Cascade Mountains of the Pacific Northwest is basalt.

Basalt

Other times, magma cools slowly and hardens below the surface. Earthquakes and other changes in Earth's crust might bring these igneous rocks to the surface years later. The granite you studied cooled below Earth's surface. The Sierra Nevada in central California and the Rocky Mountains in Colorado, Wyoming, and Montana are mostly granite mountains.

Granite

Sandstone

Sedimentary Rocks

Sedimentary rocks form from bits and pieces of recycled rocks and minerals. **Sandstone** is an example of a sedimentary rock. Sandstone starts as big rocks in the mountains. Over time, the rocks crack and break into smaller pieces. This process is called weathering.

Water can cause weathering. Water freezes in cracks in rocks. It expands when it freezes and breaks the rocks apart. Tree roots also cause weathering. Roots grow into cracks in rocks and break big pieces of rock loose.

Loose rocks tumble downhill and break into smaller pieces. Pieces might end up in streams and rivers. The pieces get banged around and broken into smaller and smaller pieces. Eventually the rocks from the mountain are reduced to tiny pieces of sand.

Sand often gets deposited in the ocean and bays. Layers of sand build up. The layers of sand are called sediment. As millions of years pass, the sand gets buried under more layers of sediment. Sand particles are pressed and stuck together. The sand turns into the sedimentary rock sandstone.

Sedimentary rocks often have bits of sand and gravel you can see. Sometimes sedimentary rocks contain fossils of shells, animals, or plants. Sedimentary rocks form in layers. If the rocks are still in their natural site, you can often see the layers.

Sandstone layers

A trilobite fossil

A fern fossil

A shell fossil

Metamorphic Rocks

Meta- means change. *Morph* means shape or form. Metamorphic rocks change from one kind of rock into another kind of rock. The starting rocks can be igneous, sedimentary, or even other metamorphic rocks. The rocks change because of heat and pressure. If a rock gets buried deep in Earth's crust or touches hot lava, it will change into metamorphic rock.

Heat and pressure can turn sandstone into quartzite. Limestone can become marble. Shale can change into slate. Heat and pressure can turn granite into gneiss (pronounced "nice").

The Rock Cycle

Metamorphic rocks aren't the only rocks that can change. Over time, any kind of rock can change into any other kind of rock. The changes from igneous to sedimentary to metamorphic and back to igneous are the **rock cycle**.

For example, a piece of igneous granite might weather into sediments. The sediments can end up in a layer with other sediments. After a long time, the sediments might change into sedimentary sandstone.

The sandstone could get heated by a lava flow or buried under other sediments. The heat and pressure might change the sandstone into metamorphic quartzite. And finally, the quartzite might be carried down into Earth's mantle where it will melt. After millions of years, the rock material might come back in a new piece of igneous granite.

Rocks don't all follow this path through the rock cycle. The important thing to remember is that all rocks change. Any rock can change into any other kind of rock. Study the rock cycle illustration to see how.

It is even possible for a rock to re-form as the same kind of rock. For example, sandstone might weather into sand. The sand could pile up in a bay. After millions of years, the sand might become new sandstone.

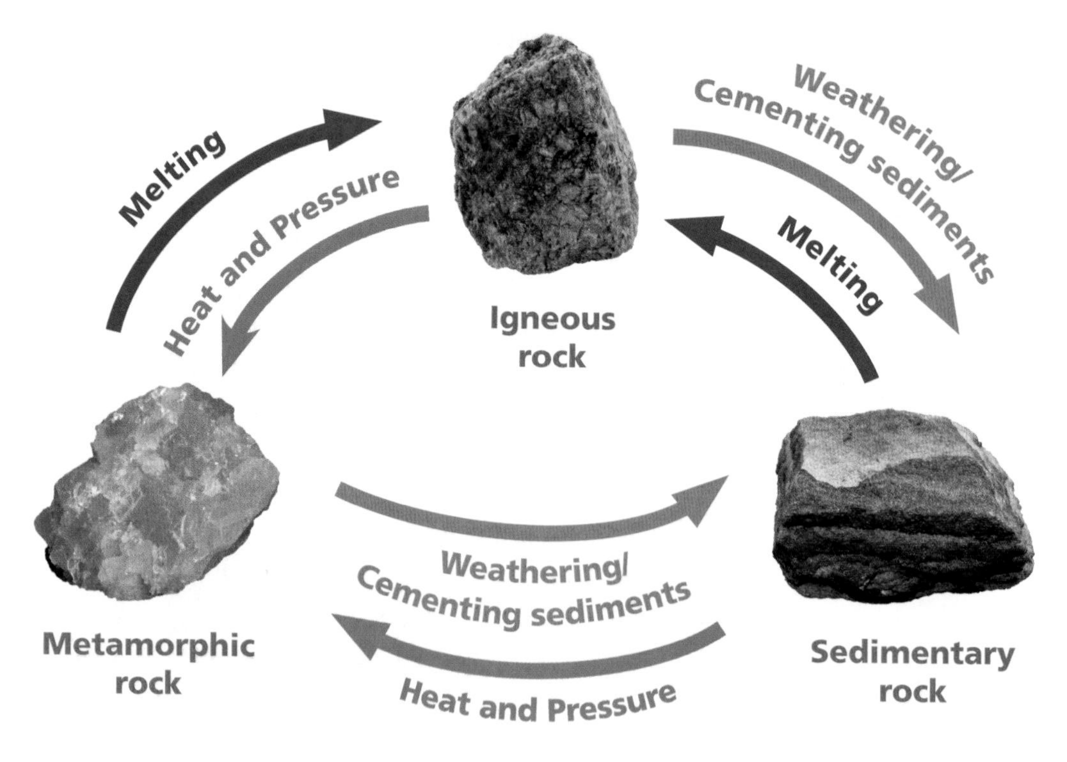

Any kind of rock can change into any other kind of rock. This is the rock cycle.

Rock Samples

This table shows several examples of sedimentary, metamorphic, and igneous rocks. How many of them have you held in your hand?

Science Safety Rules

1. Listen carefully to your teacher's instructions. Follow all directions. Ask questions if you don't know what to do.
2. Tell your teacher if you have any allergies.
3. Never put any materials in your mouth. Do not taste anything unless your teacher tells you to do so.
4. Never smell any unknown material. If your teacher tells you to smell something, wave your hand over the material to bring the smell toward your nose.
5. Do not touch your face, mouth, ears, eyes, or nose while working with chemicals, plants, or animals.
6. Always protect your eyes. Wear safety goggles when necessary. Tell your teacher if you wear contact lenses.
7. Always wash your hands with soap and warm water after handling chemicals, plants, or animals.
8. Never mix any chemicals unless your teacher tells you to do so.
9. Report all spills, accidents, and injuries to your teacher.
10. Treat animals with respect, caution, and consideration.
11. Clean up your work space after each investigation.
12. Act responsibly during all science activities.

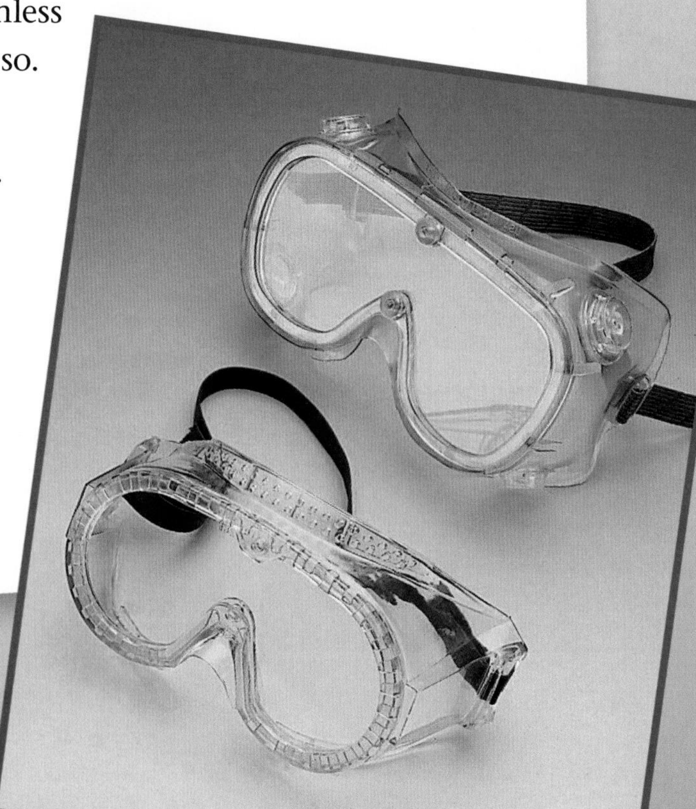

Glossary

abrasion the rubbing, grindi… bumping of rocks that cause phys… weathering

acid a substance that geologists use to identify rocks that contain calcite

calcite a common rock-forming mineral in Earth's crust

carnivore an animal that eats only animals

cast a copy of an organism, like a fossil, that is created by the minerals in a mold

chemical weathering the process by which the minerals in a rock can change due to chemicals in water and air. Chemical weathering can cause rocks to break apart.

conclusion a scientific decision based on observations, evidence, and data

concrete a mixture of gravel, sand, cement, and water

contour interval the change in elevation between any two contour lines

contour lines the curved lines in a topographic map that represent a specific elevation

core the center of Earth, made mostly of iron and nickel

crust Earth's outer layer of solid rock

decay when dead plants or animals break down into small pieces

deposition the settling of sediments

…stive system the organs and … that digest food. The …

earth… … large … resource tha… … natural … soil and water … mixes … ding

earthquake a sudd… Earth's crust along a faul…

engineer a scientist who des… accomplish a goal or solve a prob…

erosion the carrying away of weathe… earth materials by water, wind, or ice

evidence data used to support claims. Evidence is based on observations and scientific data.

fault a break in Earth's crust along which blocks of rock move past each other

flood a large amount of water flowing over land that is usually dry

fossil any remains, trace, or imprint of animal or plant life preserved in Earth's crust

fossil record all the fossils on Earth

fracture the uneven, rounded, or splintered surfaces of some minerals when they break

geologist a scientist who studies Earth, its materials, and its history

geoscientist a scientist who studies the use, distribution, and conservation of Earth's natural resources

glacier a large mass of ice mov...
over land

...ad plant

...**ite** a...hat forms when
plant...a) hardens

...**r** the numbers on contour
...etermine whether the elevation
... or falling

...**dform** a feature of the land, such as a mountain, canyon, or beach

landslide the sudden movement of earth materials down a slope

lava melted rock erupting onto Earth's surface, usually from a volcano

limestone a sedimentary rock made mostly of calcite

magma melted rock below Earth's surface

mantle the solid rock material between Earth's core and crust

marble a metamorphic rock formed when limestone is subjected to heat and pressure

metamorphic rock a rock that forms when rocks and minerals are subjected to heat and pressure

mineral an ingredient of a rock

mold a space in the sediments that fills with minerals

...**esource** living or nonliving
...s, such as soil, forests, or water,
... come from the natural environment

nutrient something that living things need to grow and stay healthy

organism any living thing

paleontologist a scientist who studies fossils

particle a very small piece or part

petrified wood the fossil remains of trees. The term means "wood turned into stone."

physical weathering the process by which rocks are broken down by breaking and banging

Portland cement a kind of cement made from limestone

predator an animal that hunts and catches other animals for food

property something that you can observe about an object or a material. Size, color, shape, texture, and smell are properties.

pumice a type of rock that forms when lava erupts from volcanoes

react to act or change in response to something

restoration putting something back to its original condition, such as building a fossil skeleton

retain to hold or continue to hold

rock a solid earth material made of two or more minerals

rock cycle the processes by which rocks change into different kinds of rocks

sandstone a sedimentary rock made of sand particles stuck together

sediment pieces of weathered rock such as sand, deposited by wind, water, and ice

sedimentary rock a rock that forms when layers of sediments get stuck together

silt rocks that are smaller than sand, but bigger than clay

soil a mix of humus, sand, silt, clay, gravel, and/or pebbles

topographic map a map that uses contour lines to show the shape and elevation of the land

transport to move or carry from one place to another

volcano an opening in Earth's crust where lava, cinders, ash, and gases come to the surface

weathering the process by which larger rocks crack and break apart over time to form smaller rocks

Index